Copyright 2021 ©

All Rights Reserved

Do Not Copy or Reproduce

Che@AdvanceWildlife.org

Created in Maui
Printed in China

Advance Wildlife Education

AdvanceWildlifeEducation.org

AWE was founded in order to build a bridge between the public and conservation organizations and nonprofits, through education, art, jewelry, and attire.

Wildlife Biologist, Founder, and Illustrator

Che Frausto

Dedicated to my parents and family

Great White Shark
Carcharodon carcharias

Fun Facts
- Great whites are the largest predatory fish on Earth.
- They can reach up to 20 feet in length.
- Great White sharks can weigh up to 3.5 tons (7,000 lbs.).
- The only two fishes that grow larger than Great Whites are whale sharks and basking sharks, both filter feeders that eat plankton.
- Of the 100-plus annual shark attacks worldwide, one-third to one-half are by great whites. However, most of these are not fatal, and new research finds that great whites, who are naturally curious, are "sample biting" then releasing their victims rather than preying on humans.
- These torpedo-shaped swimmers with powerful tails can propel them through the water at speeds of up to 15 miles per hour.
- Their mouths are lined with up to 300 serrated, triangular teeth arranged in several rows, and they have an exceptional sense of smell to detect prey.
- They have organs that can sense the tiny electromagnetic fields generated by animals.
- Though almost all fishes are cold blooded, great whites have a specialized blood vessel structure – called a countercurrent exchanger – that allows them to maintain a body temperature that is higher than the surrounding water. This adaptation provides them with a major advantage when hunting in cold water by allowing them to move more quickly and intelligently.

Diet
- Their main prey items include sea lions, seals, small toothed whales, and even sea turtles, and carrion.
- Great whites are known to take very deep dives, probably to feed on slow-moving fishes and squids in the cold waters of the deep sea.

Status
- There is no reliable data on the great white's population.
- Scientists agree that their number are decreasing due to overfishing and accidental catching in gill nets.
- They are considered a vulnerable species.

Do Not Copy or Reproduce
Copyright 2019 © Advance Wildlife Education

Great Hammerhead Shark
Sphyrna mokarran

Fun Facts
- They can live up 20 – 30 years and swim up to speeds of 25mph.
- They are usually 13 – 20 feet long and weigh 500 – 1000 lbs.
- Hammerheads are considered one of the most recently evolved groups of sharks.
- Their wide-set eyes give them a better visual range than most other sharks.
- They have sensory organs along their T- shaped head that help detect electrical fields created by prey.
- The hammerhead's increased sensitivity allows it to find its favorite meal, stingrays, which usually bury themselves under the sand.
- Their nostrils are expanded more compared with other groups of sharks and may provide hammerhead sharks with a stronger ability to locate prey and follow scents to their sources.
- Most hammerhead species are small and are considered harmless to humans.

Diet
- These sharks are hunters of the night.
- Many are bottom hunters with a preferred prey of rays, shrimps, squids, small fish, and even other shark species.
- The Great Hammerhead is feared by smaller Hammerhead species due to frequent cannibalism.

Status
- Unfortunately, many species of Hammerhead Sharks are at a high risk of extinction. Hammerhead fins are considered a delicacy in many countries. Fisherman can sell these fins for very high prices, so many times a Hammerhead is captures, has its fins removed, then is dumped back into the Ocean.

Pacific Common Thresher Shark
Alopias vulpinus

Fun Facts
- Thresher sharks grow slowly, reaching lengths up to 18 feet.
- They live a long time, between 19 and 50 years.
- They mature when they reach about 5 years old and 5 feet in length.
- They're deadly at both ends, because they've managed to weaponize their tails.
- The top halves of their scythe-like tail fins are so huge that they can be as long as the rest of the shark.
- The thresher accelerates towards a ball of fish and brakes sharply lowering its head and tail whips over its head with an average speed of 30 miles per hour.
- The fastest shark managed to whip its tail at an astonishing top speed of 80 miles per hour.
- The threshers are only successful on a third of their strikes but during these victories, they always kill several sardines at once. That's far more efficient than chasing after such quick individuals.

Diet
- Thresher sharks eat mainly anchovies, sardines, hake, mackerel, and squid.

Status
- The stock has never been assessed.

Tiger Shark
Galeocerdo cuvier

Fun Facts
- Tiger shark is the fourth largest shark and second largest predatory shark, behind only the great white.
- The largest tiger sharks can grow as much as 20-25 feet long and weigh more than 1,900 lbs.
- They live 15 or more years in the wild.
- Tiger sharks are named for the dark, vertical stripes found mainly on juveniles. As these sharks mature, the lines begin to fade and almost disappear.

Diet
- They are consummate scavengers, with excellent senses of sight and smell and a nearly limitless menu of diet items.
- They have sharp, highly serrated teeth and powerful jaws that allow them to crack the shells of sea turtles and clams.
- The stomach contents of captured tiger sharks have included stingrays, sea snakes, seals, birds, squids, and even license plates and old tires.

Status
- Tiger sharks are common in tropical and sub-tropical waters throughout the world.
- They are heavily harvested for their fins, skin, and flesh, and their livers contain high levels of vitamin A, which is processed into vitamin oil. They have extremely low repopulation rates, and therefore may be highly susceptible to fishing pressure. They are listed as near threatened throughout their range.

Shortfin Mako Shark
Isurus oxyrinchus

Fun Facts
- The shortfin mako shark is a large, predatory shark that lives in the open ocean and reaches lengths of 12 feet (3.8 m) and weights of at least 1200 pounds (545 kg).
- With top speeds of 45 miles per hour (74 kilometers per hour), the shortfin mako is the fastest shark and is one of the fastest fishes on the planet.
- It is known for its incredible leaping ability and can be observed jumping to extreme heights (out of the water) when hunting.
- Like the true tunas, the great white shark, and some other fishes, the shortfin mako shark has a specialized blood vessel structure – called a countercurrent exchanger – that allows them to maintain a body temperature that is higher than the surrounding water. This adaptation provides them with a major advantage when hunting in cold water by allowing them to move more quickly and intelligently.
- While the shortfin mako shark is one of only very few shark species known to have bitten and killed people, these events are extremely rare and likely accidental (a case of mistaken identity).

Diet
- This species feeds on a variety of prey. They are known to primarily eat bony fishes (including relatively large tunas) and squids but also eat other sharks, small marine mammals, sea turtles, and even dead organic matter.
- Shortfin makos are at the top of the pelagic food web, and adults do not have any known natural predators.

Status
- Shortfin mako sharks have a large geographical range and are found widely in tropical to temperate latitudes of all oceans. Everywhere that they live, they are either targeted commercially or captured accidentally in fisheries targeting other species. These sharks are valued for the high quality of their fins and meat. Fishers that use longline fishing gear to target swordfish, yellowfin tuna, and other tunas regularly capture shortfin mako sharks and keep them to sell commercially. Other fisheries use longline or gillnet fishing gear to specifically target these sharks. The combination of these practices is driving down populations of shortfin makos all around the world, and scientists now believe them to be vulnerable to extinction. Without increased conservation and management efforts, this species' populations will continue to decline, perhaps to a dangerous degree.

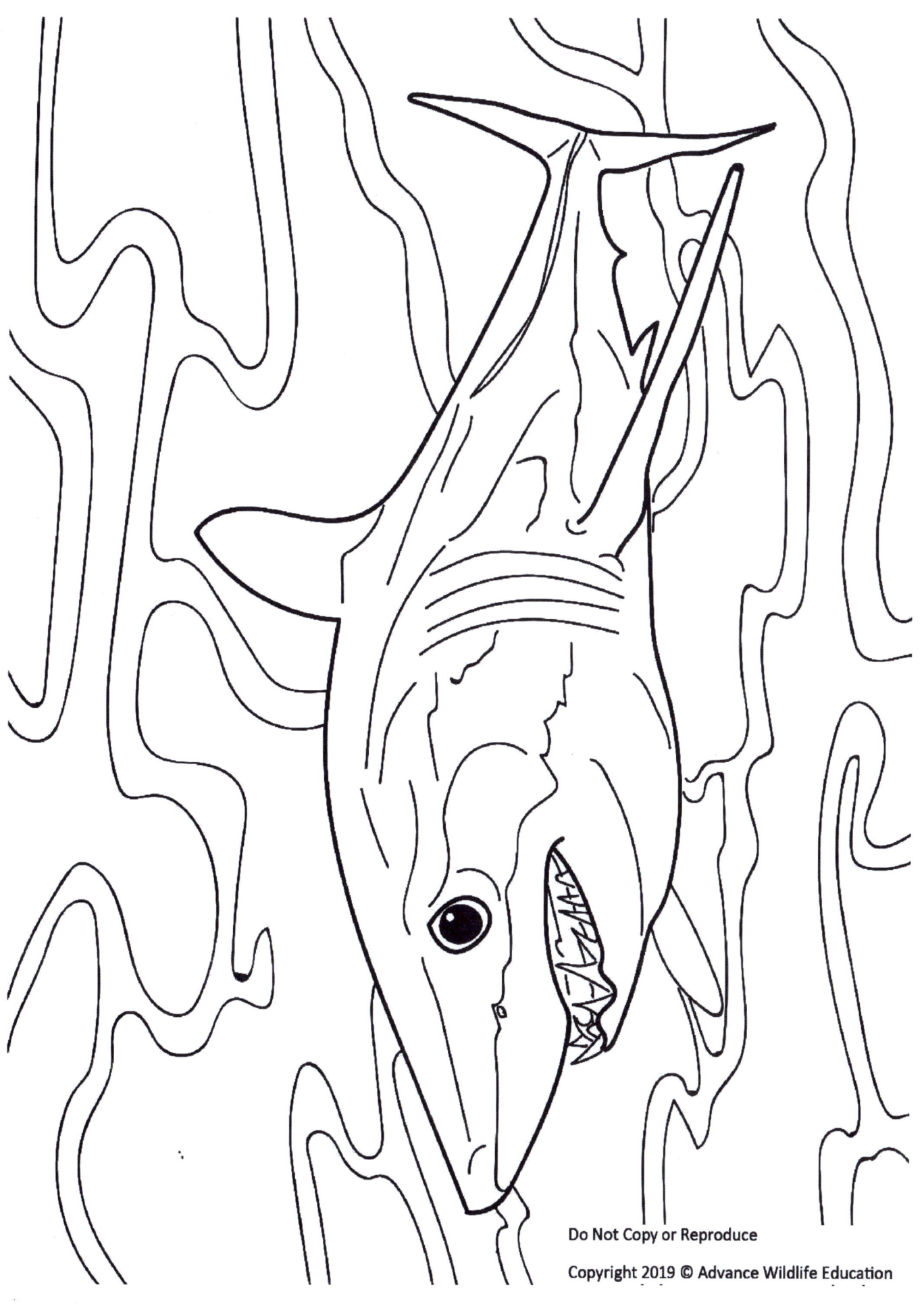

Blue Shark
Prionace glauca

Fun Facts
- The common name comes from the blue color of the skin, unique among the sharks.
- They may reach lengths of up to 9-10 feet.
- The blue shark has one of the largest geographic ranges among the sharks and was historically one of the most (if not *the* most) common pelagic sharks in the world.
- Females give live birth, and litters are known to rarely reach sizes of more than 100 pups.
- They can dive up to 350 meters (1150 feet) deep.

Diet
- Blue sharks specialize in relatively small prey including octopus, squid, mackerel, tunas, lobsters, crabs, small sharks and occasionally seabirds.

Status
- Near threatened with extinction.
- Blue sharks are a target of fisheries in some areas and a common accidentally caught species in gillnet and longline fisheries targeting other species.
- Its fins are considered highly valuable, and blue sharks may be the target of illegal 'shark finning' operations, where the fins are cut off and kept, while the rest of the shark is wasted.

Leopard Shark
Triakis semifasciata

Fun Facts
- They are generally 50 and 60 inches long but can grow up to 7 feet long.
- Female leopard sharks can produce litters of 4 to 33 pups.
- Believed to be a nocturnal hunter, they spend most of the day lazily swimming and resting on the bottom, becoming active at night when they hunt for sleeping fish.

Diet
- Their diet includes invertebrates such as crabs, shrimp, octopi, fat innkeeper worms, clam siphons, and fish.

Status
- Due to the relatively late age of first reproduction, the slow growth rate, and the low reproduction rate, the leopard shark is potentially threatened by over-fishing. They are considered vulnerable.

Bull Shark
Carcharhinus leucas

Fun Facts
- They weigh 200- 500 lbs and are 7 – 11.5 ft long.
- Their average lifespan in the wild is 16 years.
- Bull sharks are aggressive, common, and usually live near high-population areas like tropical shorelines.
- These large stout sharks are found in both salt and fresh water.
- They have been recorded in rivers hundreds of miles from the sea.
- They are not bothered by brackish and freshwater, and even venture far inland via rivers and tributaries. Because of these characteristics, many experts consider bull sharks to be the most dangerous sharks in the world.
- Bull sharks get their name from their short, blunt snout, as well as their tendency to head-butt their prey before attacking.
-

Diet
- Fast, agile predators, they will eat almost anything they see, including fish, dolphins, and even other sharks.

Status
- Bull sharks are fished widely for their meat, hides, and oils, and their numbers are likely shrinking.
- Though the bull shark is not a targeted species in most commercial fisheries, it is regularly captured on bottom longline gear.
- IUCN Red List Status: Near Threatened

Whitetip Reef Shark
Triaenodon obesus

Fun Facts

- This shark is a very slim species. At most, it grows rarely over 5.25 ft (1.6 m) and weighs up to 20 kilograms (44 pounds).
- Long, slender bodies allow whitetips to maneuver through crevices and caves in their coral reef habitats.
- They have a wide range of distributions in Indo-Pacific, Central Pacific and Eastern Pacific Oceans, also off Central America, usually in shallow areas close to shore on or near coral reefs
- Whitetips forage at night and spend their days resting in reef caves.
- They're not territorial—many sharks crowd into a cave, usually stacking themselves atop each other.
- On occasion, a whitetip might approach a diver out of curiosity but is not considered dangerous to humans.
- Sharks play a crucial role as apex predators in keeping marine ecosystems in balance and removing sick, injured and diseased animals. Therefore, they are vital to having a clean and healthy ocean.

Diet

- Whitetip reef sharks are active nocturnal predators with a diet that includes bony fish, crustaceans and octopus.

Status

- Whitetip shark populations aren't in danger, but they live in shallow water in a restricted habitat where fisheries can catch them easily using gill nets and longlines. Additionally, these sharks mature late and have small litters. With increased fishing pressures, this species may become threatened.

Lemon Shark
Negaprion brevirostris

Fun Facts

- The lemon shark is easily recognized for its yellow-brown to olive color- an ideal camouflage against the sandy in-shore areas they prefer to forage in.
- The lemon shark is commonly found in subtropical shallow water to depths of 300 feet (90 m) around coral reefs, mangroves, enclosed bays, sounds and river mouths.
- Lemon sharks are one of the larger species of sharks, commonly obtaining lengths between 250-300 cms (8 to 10 feet) and can weigh up to 250 kgs (551 lbs).
- Females and males reach sexual maturity around 12-13 years of age.
- Maximum life expectancy is about 27 years.

Diet

- The lemon shark is commonly found swimming over sandy or muddy bottoms and eats a diet consisting mainly of bony fish and crustaceans. Catfish, mullet, jacks, croakers, porcupine fish, cowfish, guitarfish, stingrays, eagle rays, crabs and crayfish make up the majority of their diet. In addition, this species will eat sea birds and smaller sharks.

Status

- IUCN Red List Status: Near Threatened
- The lemon shark is targeted by commercial and recreational fisheries along the US Atlantic Ocean, Caribbean, and in the eastern Pacific Ocean. The US bottom longline fishery commonly targets this species and it is also caught as by-catch in both pelagic and gillnet fisheries. The meat is consumed in Central America, South America, and the US. Their fins are highly prized and exported to Asia for shark fin soup and their skin is used for leather.

Whale Shark
Rhincodon typus

Fun Facts

- As the largest fish in the sea, reaching lengths of 40 feet or more.
- Preferring warm waters, whale sharks populate all tropical seas.
- They feed on plankton and travel large distances to find enough food to sustain their huge size, and to reproduce.
- Adults are often found feeding at the surface but may dive to 1000m.
- They weigh around 11 tons (22,000 lbs).
- Unlike other large sharks, which give birth to a small number of very large babies, whale sharks give birth to hundreds of very small babies (approximately 20 inches/45 cm).

Diet

- Whale sharks eat tiny plankton and fish eggs, which they filter feed as they swim slowly along with their giant mouths wide open. They are one of only three species of filter feeding sharks.

Status

- Although massive, whale sharks are docile fish and sometimes allow swimmers to hitch a ride. They are currently listed as a vulnerable species; however, they continue to be hunted in parts of Asia, such as the Philippines.

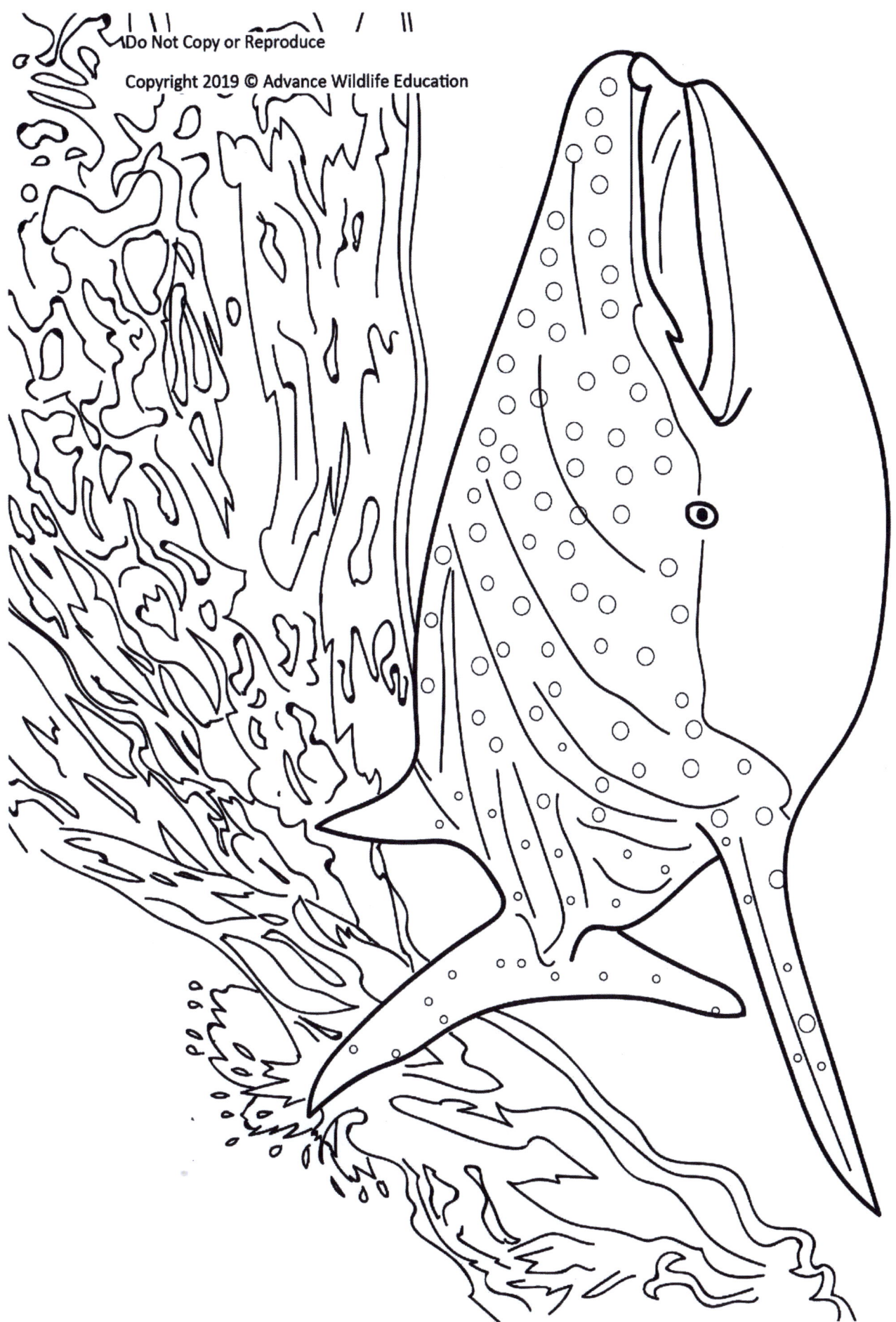

Basking Shark
Cetorhinus maximus

Fun Facts
- This slow-moving migratory shark is the second largest fish, growing as long as 32 feet and weighing over 5 tons.
- Basking sharks are passive and no danger to humans in general.
- Its English name "basking shark," which means "taking the sun," comes from its habit of swimming very close to the surface with the dorsal fin out of the water.
- Interestingly, many tales of sea serpents and monsters have originated from sightings of basking sharks.
- The basking shark has a very large liver that accounts for up to 25% of its body weight.

Diet
- It is often sighted swimming close to the surface, huge mouth open, filtering 2,000 tons of seawater per hour over its complicated gills to scoop up zooplankton.

Status
- In the past, basking sharks were hunted worldwide for their oil, meat, fins, and vitamin rich livers. Today, most fishing has ceased except in China and Japan.
- They have a lengthy maturation time, slow growth rate and a long gestation period. These factors combined with an already depleted population in many areas have prompted many countries to establish laws to protect the basking shark from further exploitation.
- The North East Atlantic population are classed as Endangered on the IUCN Red List.

Megamouth Shark
Megachasma pelagios

Fun Facts
- The megamouth shark is a rare shark and a large species, reaching weights of 2700 pounds (1215 kg).
- However, it is the smallest of the three species of filter-feeding sharks, behind the whale shark and the basking shark.
- Dark gray brown on top and light gray to white below, they are thought to grow to 17 feet long.
- The megamouth shark gets its name from the remarkably large, circular mouth. On an individual approximately 16 feet in length (5 m), the mouth is approximately four feet across (1.3 m).
- Megamouth sharks live from near the surface to as deep as 15,000 feet (4600 m).
- Scientists believe megamouth sharks only come near the surface at night and spend most of their lives in the dark.
- Though it is one of the largest sharks in the world, the megamouth shark was only discovered by scientists in 1976.
- This species has only been observed in the wild a few times, and less than 60 individuals are known by scientists to ever be captured or observed.
- The megamouth has approximately fifty rows of very small and relatively numerous teeth on each jaw, but only three rows are functional.
- The only confirmed register of a megamouth predator is an isolated event of sperm whales (Physeter macrocephalus) attacking this shark.

Diet
- They are filter feeders and swim with their mouths constantly wide open in order to filter out their preferred planktonic prey. The inside of their mouths are covered with light producing organs that may be used to attract pelagic crustaceans and other potential prey.

Status
- It is likely naturally very rare, but scientists do not believe that they have sufficient knowledge of this species to determine its conservation status.

Zebra Shark
Stegostoma fasciatum

Fun Facts

- The maximum reported size of a male zebra shark is 7.7 feet (2.35 m) total length and the maximum size of a female is 7.6 feet (2.33 m) total length.
- The zebra shark lives on sand, rubble, and coral bottoms at depths from 0-207 feet (0-63 m).
- The lifespan of the zebra shark is believed to be 25-30 years in the wild.
- Due to its sluggish behavior during daylight hours, it is believed to be a nocturnal species. It is often observed sitting on the bottom near coral reefs.
- This small shark has been recorded in marine and brackish waters as well as in freshwater habitats.
- It has a rounded snout and small eyes, rounded pectoral fins, small dorsal fins, and a small mouth with teeth made for crunching mollusks.
- They are considered harmless to humans and are popular attractions for dive tourism and public aquariums.

Diet

- This small shark feeds primarily on mollusks as well as small bony fishes that it sucks out of the sand. Other prey items include crabs and shrimp. It also swims slowly and squirms into narrow crevices and channels in reefs in search of prey.

Status

- Zebra sharks are susceptible to declines in population due to inshore fishery activities and coral reef habitat loss because of its limited habitat preference and geological distribution. It is listed as "Vulnerable" by the World Conservation Union (IUCN) throughout most of its range.

Goblin Shark
Mitsukurina owstoni

Fun Facts

- Spotted mostly off the coast of Japan, they're named for their likeness to mythical goblins that appear in Japanese folklore.
- This bottom-dwelling shark is pinkish grey and can grow up to 12 feet long and weigh up to 460 pounds.
- It has a long, prominent snout covered with special sensing organs that help it to sense electric fields in the deep, dark water.
- The shark can thrust its jaw three inches out of its mouth. (The jaw is connected to three-inch-long flaps of skin that can unfold from its snout).
- Most specimens have been observed near continental slopes between 885 feet (270m) and 3149 feet (960m) deep. It has been found in waters up to 4265 feet (1,300m) deep and in waters as shallow as 311 feet (95m) to 449 feet (137m).

Diet

- It is thought to mostly eat soft prey like shrimp, small fish, octopus, and squid, which it catches by quickly projecting its jaw forward and pulling prey into its mouth.

Status

- Least Concern
- It is fished only as a bycatch of Deepwater trawl, longlines and deep-set gill nets.

Cookiecutter Shark
Isistius brasiliensis

Fun Facts
- It attaches itself to the prey and uses its serrated bottom teeth to cut out a perfectly circular chunk of flesh.
- The cookiecutter shark is named after the cookie-shaped wounds that it leaves on the bodies of its prey items.
- This small, cigar-shaped shark is dark brown on top and light on the underside, with a darker band around its neck.
- Found in deep water at depths below 3,281 feet (1000 m) during the day, cookiecutter sharks migrate vertically to surface waters at night to feed.
- Cookiecutter sharks have 30-37 small, erect teeth in the upper jaw and 25-31 larger triangular teeth in the lower jaw.
- Male cookiecutter sharks grow to a maximum of 16.5 inches (42 cm) total length (TL) while females reach 22 inches (56 cm)

Diet
- Common prey items include large fishes such as marlin, wahoo, dolphin, tuna, sharks, and stingrays as well as marine mammals including seals, whales, and dolphins.

Frilled Shark
Chlamydoselachus anguineus

Fun Facts
- The frilled shark is a strange, prehistoric-looking shark that lives in the open ocean and spends much of its time in deep, dark waters far below the sea surface.
- Its long, cylindrical body reaches lengths of nearly 7 feet (2 m), and its fins are placed far back on the body.
- The frilled shark gets its name from the frilly appearance of its gill slits.

Diet
- The preferred prey of the frilled shark is squid, and they have several rows of long teeth, each with three long points, that are perfect for snagging the soft bodies of this prey. Though they specialize on squids, frilled sharks are known to eat a variety of fishes and also other sharks.

Status
- Frilled sharks are only very rarely encountered in the wild, so little is known about their ecology and population.

Common Eagle Ray
Myliobatis aquila

Fun Facts
- This species lives at depths of 3-985 feet (1-300 m), the common eagle ray resides in tropical to warm temperate waters along coastlines in shallow lagoons, bays and estuaries.
- This ray often schools in small groups in bays, shallow lagoons, and estuaries, stirring up crustaceans and mollusks that it crushes with its plated teeth.
- The maximum reported size of the common eagle ray is a disc width of 72 inches (183 cm) and the maximum reported weight is 32 pounds (14.5 kg).

Diet
- Common eagle rays feed on crustaceans and mollusks by dislodging them from the sediments. They are also known to feed on fishes.

Status
- The common eagle ray is not listed as threatened or endangered by the World Conservation Union (IUCN).

Shovelnose Guitarfish
Rhinobatos productus

Fun Facts
- Compressed from belly to back, guitarfish bodies are attuned to life on the sand.
- They live on sandy seafloors in bays, seagrass beds and estuaries, and usually in less than 40 feet (12 m) of water.
- This ancient ray has been playing it flat for over 100 million years.
- The guitarfishes are a group of skates (as opposed to stingrays). They do not have barbs or "stingers" like some other rays, and they are totally harmless to people.

Diet
- Guitarfish lie in ambush buried in the sand with only their eyes sticking out, waiting for an unwary crab or flatfish to wander by.
- At night, they leave the sand to actively cruise the seafloor to feed on crabs, worms, clams and, perhaps, fishes.

Status
- In some areas (particularly in northern Mexico), their numbers have been depleted significantly, and scientists now believe that the species is near threatened with

Giant Manta Ray
Manta birostris

Fun Facts
- The giant manta ray is the world's largest ray with a wingspan of up to 29 feet.
- Giant manta rays are slow-growing, migratory animals with small, highly fragmented populations that are sparsely distributed across the world.
- There are reports of giant mantas living to at least 40 years, but little is known about their growth and development.
- The largest specimens of the manta weigh up to 3,000 pounds (1350 kg).

Diet
- They are filter feeders and eat large quantities of zooplankton.
- They constantly swim along with their large mouths open, filtering plankton and other small food from the water. To aid in this strategy, giant mantas have specialized flaps, known as cephalic lobes, which help direct more water and food into their mouths.

Status
- The main threat to the giant manta ray is commercial fishing, with the species both targeted and caught as bycatch in a number of global fisheries throughout its range. Manta rays are particularly valued for their gill rakers, which are traded internationally. In 2018, NOAA Fisheries listed the species as threatened under the Endangered Species Act.

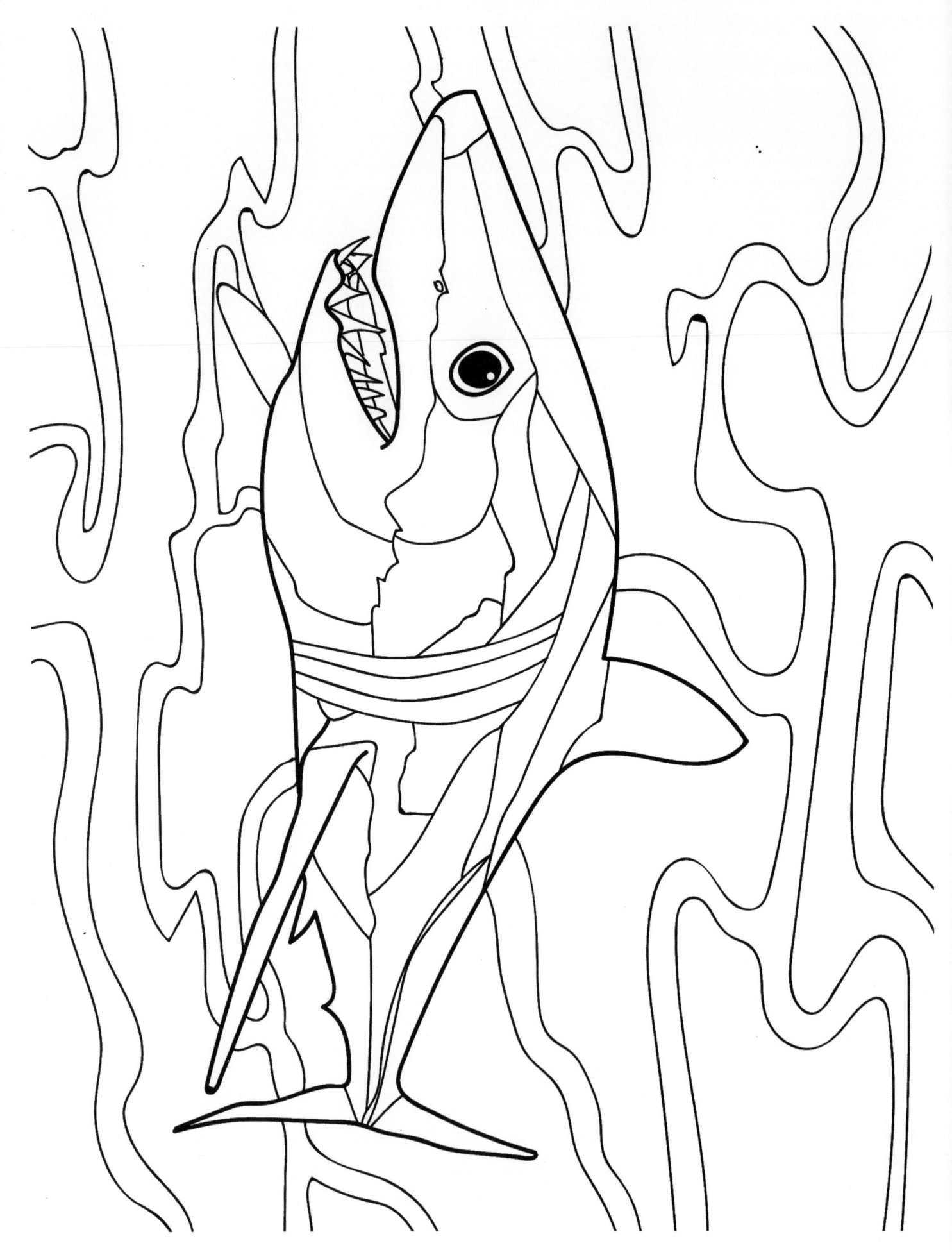

The deadliest animals.
Average annual animal-caused fatalities in the U.S., 2001 to 2013

Sharks kill 1 person per year.

Alligators kill 1 person per year.

Bears kill 1 person per year.

Venomous snakes and lizards kill 6 people per year.

Spiders kill 7 people per year.

Non-venomous arthropods kill 9 people per year.

Cows kill 20 people per year.

Dogs kill 28 people per year.

Other mammals kill 52 people per year.

Bees, wasps and hornets kill 58 people per year.

WAPO.ST/WONKBLOG
Sources: CDC reports, CDC WONDER database, Wikipedia, Florida Museum of Natural History

As apex predators, sharks play an important role in the ecosystem by maintaining the species below them in the food chain and serving as an indicator for ocean health. They help remove the weak and the sick as well as keeping the balance with competitors helping to ensure species diversity.

What can you do to help wildlife?

- Reduce use of plastics
- Reusable drinking containers
- Using reef safe sunscreen
- Waiting 30 min after applying sunscreen to go into water
- Not standing or walking on coral reefs
- Keeping a respectful distance from wildlife
- Letting endangered sea turtles and monk seals rest peacefully on the beach
- Staying out of seabird burrowing areas because they are extremely fragile and could collapse with chicks inside
- Using wildlife friendly yellow and low frequency lighting outside to lower the risk of confusing seabirds and sea turtles that use the moon and stars to navigate
- Facing lights downward and using shields making them dark sky friendly
- Only have lights on when they are in use
- Make sure there is no standing water around your property to decrease mosquito breeding
- Do not feed stray cats
- Keeping domestic pet cats indoors
- Putting a bell on your cat helps to warn other wildlife but keeping your cat inside is better
- Keeping dogs on leashes especially near seabird beach breeding grounds
- Help honeybees by planting flowers that supply nectar and pollen throughout the season
- Plant native species
- Avoid using any poisons or chemicals outside
- Avoid using chemical fertilizers in your garden or yard
- Get out and vote for politicians that support and protect wildlife
- Join a conservation organization
- Consider a career in conservation/biology/environmental studies
- Seek out internships with environmental organizations
- Attend a beach cleanup
- Pick up trash every time you go to the beach
- Try to reduce carbon footprint
- Don't put hazardous substances down the drain or in trash
- Use cloth not paper napkins
- Recycle everything you can
- Don't leave water running
- Wash laundry using cold water instead of warm

Unsustainable Fishing

Demand for seafood and advances in technology have led to fishing practices that are depleting fish and shellfish populations around the world.

Fishers remove more than 77 billion kilograms (170 billion pounds) of wildlife from the sea each year. Scientists fear that continuing to fish at this rate may soon result in a collapse of the world's fisheries. In order to continue relying on the ocean as an important food source, economists and conservationists say we will need to employ sustainable fishing practices.

Consider the example of the bluefin tuna. This fish is one of the largest and fastest on Earth. It is known for its delicious meat, which is often enjoyed raw, as sushi. Demand for this particular fish has resulted in very high prices at markets and has threatened its population. Today's spawning population of bluefin tuna is estimated at 21 to 29 percent of its population in 1970.

Since about that time, commercial fishers have caught bluefin tuna using purse seining and longlining.

Purse seine fishing uses a net to herd fish together and then envelop them by pulling the net's drawstring. The net can scoop up many fish at a time, and is typically used to catch schooling fish or those that come together to spawn.

Longlining is a type of fishing in which a very long line—up to 100 kilometers (62 miles)—is set and dragged behind a boat. These lines have thousands of baited hooks attached to smaller lines stretching downward.

Both purse seining and longlining are efficient fishing methods. These techniques can catch hundreds or thousands of fish at a time but are extremely harmful to bycatch species such as seabirds, dolphins, sea turtles, swordfish, and many more.

Catching so many fish at a time can result in an immediate payoff for fishers. Fishing this way consistently, however, leaves few fish of a species left in the ocean. If a fish population is small, it cannot easily replenish itself through reproduction.

Taking wildlife from the sea faster than populations can reproduce is known as overfishing.

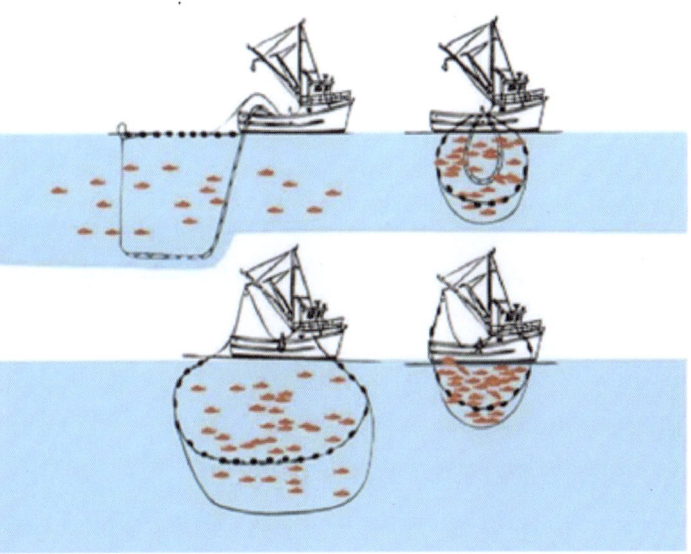

Sustainable Fishing

There are ways to fish sustainably, allowing us to enjoy seafood while ensuring that populations remain for the future.

Traditional Polynesian cultures of the South Pacific have also always relied on the ocean's resources. The most common historical fishing practices were hook and line, spearfishing, and cast nets.

Hooks constructed of bone, shell, or stone were designed to catch specific species. Fishers would also craft 2-meter (6-foot) spears. They would dive underwater or spear fish from above, again targeting specific animals. Cast nets were used by fishers working individually or in groups. The nets could be cast from shore or canoes, catching groups of fish. All of these methods targeted fish needed for fishers' families and local communities.

Some of these sustainable fishing practices are still used today. Native Hawaiians practice cast-net fishing and spearfishing.

Modern spearfishing is practiced all over the world, including in South America, Africa, Australia, and Asia. In many cases, spearguns are now used to propel the spear underwater. Spearfishing is a popular recreational activity in some areas of the United States, including Florida and Hawaii. This fishing method is considered sustainable because it targets one fish at a time and results in very little bycatch.

Rod-and-reel fishing is a modern version of traditional hook-and-line. Rods and reels come in different shapes and sizes, allowing recreational and commercial fishers to target a wide variety of fish species in both freshwater and saltwater.

Rod-and-reel fishing results in less bycatch because non-targeted species can be released immediately. Additionally, only one fish is caught at a time, preventing overfishing. For commercial fishers, rod-and-reel-fishing is a more sustainable alternative to longlining.

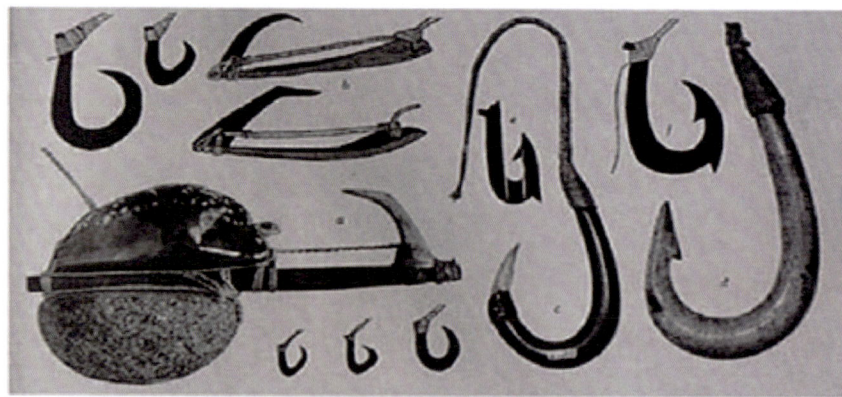

PRACTICE CATCH AND RELEASE

Great White Shark

Great Hammerhead

Thresher Shark

Tiger Shark

Shortfin Mako Shark

Blue Shark

Leopard Shark

Bull Shark

Whitetip Reef Shark

Lemon Shark

Whale Shark

Basking Shark

Megamouth Shark

Zebra Shark

Goblin Shark

Cookie Cutter Shark

Frilled Shark

Common Eagle Ray

Shovelnose Guitarfish   Manta Ray

Please leave a Google Review
Thank you for your support!

Google
Advance Wildlife Education Maui

Click Reviews

Rate by Stars

Photos: Shutterstock

Sources
Noaa.gov
NPS.gov
USFW.gov
Nationalgeographic.com
Worldwildlife.org

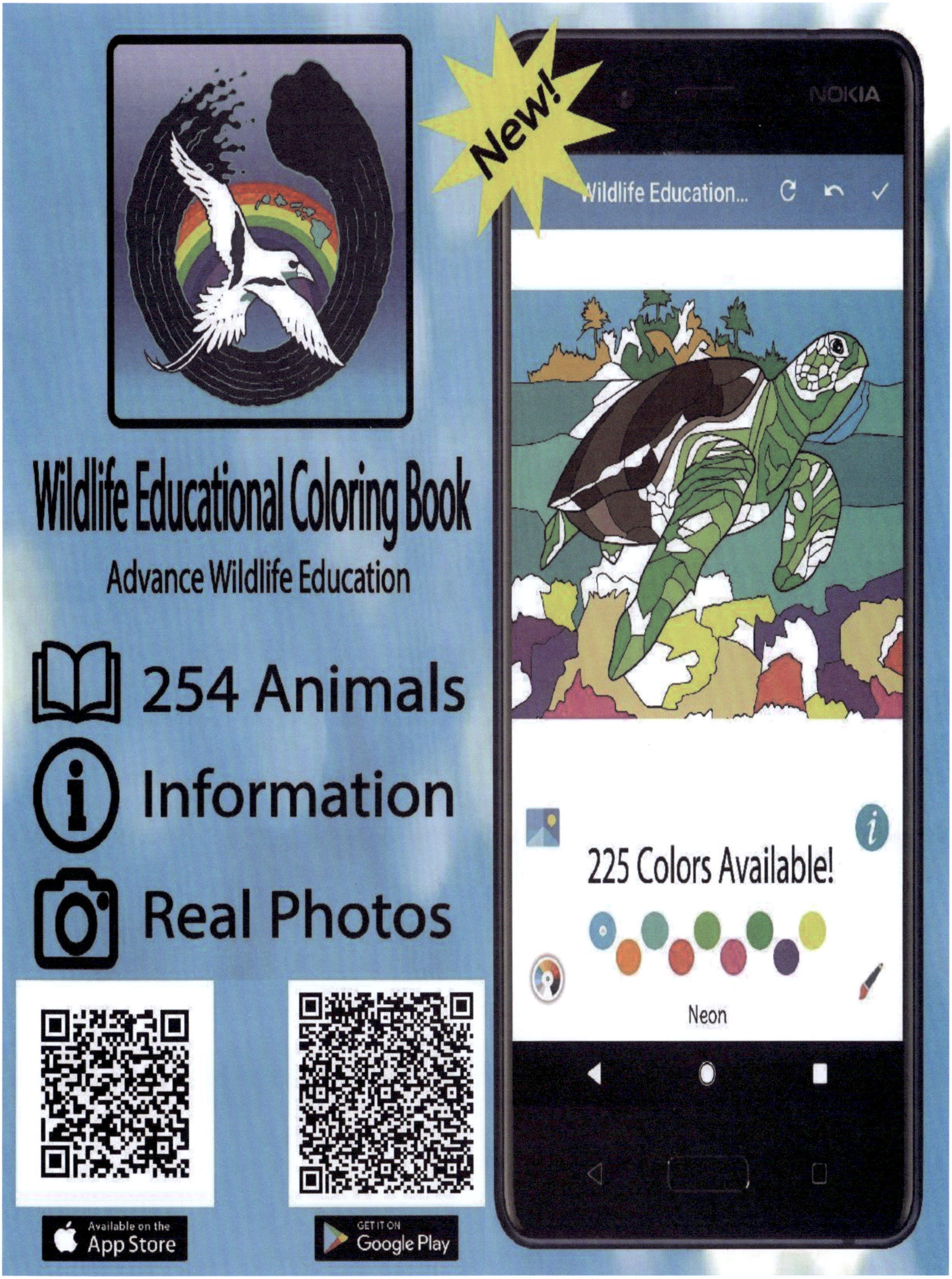

Wildlife Educational Coloring Book
Advance Wildlife Education

 **254 Animals**

 **Information**

 **Real Photos**

1,970	23.6 K
Posts	Followers

Advance Wildlife Education
🪶 Wildlife Biologist @che_frausto
🌴 Founded on Maui, Hawai'i
🌎 Daily Wildlife Education Posts
⬇️ Wildlife Educational Coloring Books
linktr.ee/advancewildlifeeducation
Paia, Hawaii 96779

Che　　Products　　Top Posts

Daily Fun Wildlife Facts
Follow us on Instagram and Facebook

@Advance_Wildlife_Education

Advance Wildlife Education in the News and Community

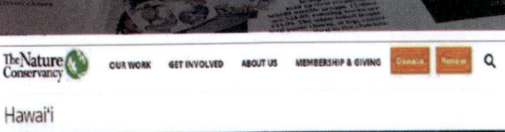
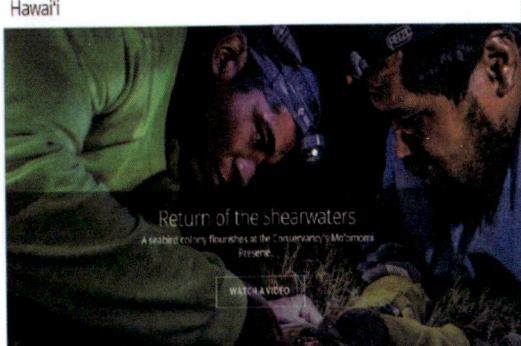

Like the Books?
Collect Them All
Order Online
AdvanceWildlifeEducation.org

Books · Stickers · Tattoos · Bamboo Tshirts · Bags · Wrap Bracelets

  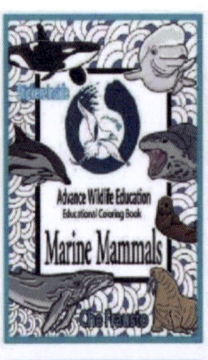 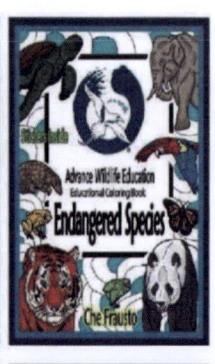